Who Am I?
River Animals

Written by Read With You
Center for Language Research and Development

Published by Read With You Publishing

Designed by Read With You Center for Language Research and Development

ISBN-13: 978-1-944710-80-4
ISBN-10: 1-944710-80-9

Printed in the United States of America.

My eyes can glow.

Who am I?

alligator

I'm pink because of the food I eat.

Who am I?

flamingo

My fur traps air to keep me warm.

Who am I?

river otter

I can hold up to three gallons of water in my pouch.

Who am I?

pelican

I catch my prey with my tongue.

Who am I?

frog

I can sleep while standing on one leg.

Who am I?

swan

I'm good at building dams.

Who am I?

beaver

I don’t have scales during the first month of my life.

Who am I?

trout

Alligator

Flamingo

River Otter

Pelican

I am an alligator.

I am a flamingo.

I am a river otter.

I am a pelican.

Frog

Swan

Beaver

Trout

I am a frog.

I am a swan.

I am a beaver.

I am a trout.

We are river animals.

A river is a flowing, moving stream of water.

Made in United States
Orlando, FL
07 November 2026

90337535R00017